Math Olympiad Contests Preparation For Grades 4-8

Competition Level Math for Middle School Students to Win MathCON, AMC-8, MATHCOUNTS, and Math Leagues

ARNOLD BETHUNE

Contents

PREFACE

You have arrived to the world of mathematical competitions.

The field of mathematics encompasses a vast universe of difficulties, puzzles, and logical thinking; it is not simply a collection of numbers and equations. This book is your guide into the exciting world of Math Olympiads, and it was written specifically for students in grades 4 through 8 who are enrolled in middle school. The purpose of this book is to prepare you for some of the most prestigious math competitions, such as MathCON, AMC-8, MATHCOUNTS, and other Math Leagues. Whether you are just starting out in the world of math competitions or are trying to improve your skills, this book will help you become ready for these events.

A Brief Introduction to the Math Olympiads

When it comes to mathematics, the Math Olympiads are more than just competitions; they are portals to learning the complexity and beauty of mathematics. Participating in these competitions will test your ability to solve problems, inspire you to think logically, and motivate you to develop a passion for mathematics. Mathematical Olympiads have been extremely helpful in locating and cultivating young mathematical talent all across the world. In this section, we delve into the history of Math Olympiads, discussing their significance on a global scale as well as the impact they have had on the field of mathematics education.

In middle school, the significance of mathematical competitions is 1.2

During the middle school years, taking part in mathematical competitions can be a life-changing experience. Not only does it help you improve your mathematics skills, but it also assists you in increasing your ability to think critically, manage your time effectively, and persevere. MathCON, AMC-8, and MATHCOUNTS are examples of competitions that are intended to test your mathematical knowledge and take you beyond the boundaries of what is often taught in schools. The purpose of this section is to examine the ways in which participating in these competitions might contribute to your academic and personal development, so laying a solid groundwork for future mathematical undertakings.

1.3 Instructions on How to Make Use of This Book

The objective of this book is to provide you with a comprehensive journey through the process of preparing for the Math Olympiad. The course begins with the fundamentals of mathematics and then gradually progresses to more difficult topics and particular methods for mathematical competitions. Learning is reinforced by the use of theory, examples, and practice problems that are included in each chapter. For the purpose of providing you with an insider's perspective, we have also included suggestions and guidance from veterans of the Olympiad as well as experts.

A methodical approach is what we suggest you do in order to get the most out of this book:

Proceed from the very beginning: To ensure that you have a strong foundation, it is beneficial to begin from the very beginning, even if you are already familiar with some of the basic topics included in the course.

The key to success in Math Olympiads is consistent practice, therefore make sure you practise on a regular basis. Create a timetable and be sure you stick to it.

Learn from blunders: When you solve practice problems, take time to comprehend your blunders. This is where genuine learning happens.

Use Resources Wisely: Apart from this book, use the recommended resources and further reading to deepen your understanding.

Stay Curious: Always be open to learning new concepts and strategies in mathematics.

1.4.1 The Road That Lies Ahead

The process of getting ready for the Math Olympiads is a journey that is full of both trials and victories. We hope that by reading this book, you will find that your travel is less stressful and more pleasurable. For the purpose of gradually enhancing both your understanding and your abilities, each chapter is intended to build upon the one that came before it. We have taken great effort in selecting the practice problems and solutions that will accurately reflect the nature and level of complexity of the questions that you will be asked during the actual competitions.

Participating Actively in the Mathematical Community

Being a part of a community that is interested in mathematics in the same way that you are may be an extremely motivating experience. Your participation in math clubs, participation in online forums, and networking with other students who are also preparing for these competitions are all things that we strongly recommend you do. Your educational experience can be considerably improved by engaging in activities such as working together, discussing issues, and exchanging ideas.

Words of encouragement to conclude, number 1.6

Always keep in mind that the purpose of competing in Math Olympiads is not simply to win, but rather to take pleasure in the process of learning and finding solutions to problems. Your progress toward becoming a better mathematician will be accelerated by each and every issue that you solve and each and every obstacle that you conquer. Therefore, set out on this trip with a positive attitude, a strong commitment, and an open mind.

While you are flipping through these pages, we hope that you will not only find lessons in mathematics, but also lessons in tenacity, problem-solving, and the joy of learning. Hello, and welcome to the marvelous world of the Math Olympiads!

1

Fundamentals of Mathematics

As we progress through this chapter, we will delve into the fundamental notions of mathematics, which are the foundation upon which any mathematical competition is built. It is absolutely necessary to have a firm understanding of these fundamental notions in order to be able to successfully handle more challenging issues in Math Olympiads. Each lesson is designed to help you strengthen your fundamental mathematics abilities, which will ultimately allow you to build a strong foundation upon which you can construct a more advanced theoretical understanding.

Introduction to Arithmetic: the Basics

Arithmetic serves as the main building block upon which the majority of mathematical computations are constructed. Within this section, we will once more go over the essential operations that are involved in arithmetic. Calculations such as addition, subtraction, multiplication, and division are included in these procedures. The processes in question are not merely concerned with numerical values; rather, they require an understanding of patterns, connections, and strategies for resolving situations that are troublesome. In this lesson, we are going to look into a number of different approaches to solving word problems, performing mental calculations, and making estimates in a short amount of time. As an additional benefit, you will be provided with an introduction to the fundamental concepts of prime numbers, factors, and multiples, all of which are featured in this section of the discussion of number theory.

2.2 An Exposition of the Concepts Employed in Algebra

Within the realm of mathematics, the language that we employ to express mathematical concepts is referred to as algebra. It is not sufficient to simply uncover the unknowns; rather, it is essential to have an understanding of the relationships that exist between the variables already in existence. Among the essential ideas that are discussed in this section of the algebraic language are expressions, equations, and inequalities. These are only some of the concepts that are explored. We will cover a variety of topics, some of which include the understanding of the properties of algebraic expressions, the ability to simplify expressions, and the ability to solve linear equations. Having this fundamental knowledge is very necessary in order to be able to tackle more challenging topics in algebra, such as quadratic equations and polynomial functions.

2.3 Acquiring Geometrical Knowledge and Understanding

The field of study known as geometry is concerned with the examination of all shapes, sizes, and the properties of space. There is no question that you absolutely need to have a strong understanding of geometry in order to be successful in Math Olympiads. The purpose of this section is to provide an introduction to the principles of geometric figures, which include points, lines, angles, and shapes. Throughout the course of this lesson, we will study the features of a wide variety of geometric figures, such as circles, triangles, quadrilaterals, and others. Additionally, there will be a debate on a variety of issues, including the Pythagorean theorem, congruence, and likeness. In order to successfully conquer challenging geometry issues, it is essential necessary to have a good understanding of these concepts.

The Principles of Number Theory, the Foundational Elements

In the field of mathematics, number theory is an exciting subject that focuses on the features and qualities of numbers, particularly integers, as well as the connections between them. This part will provide a concise introduction to the fundamentals of number theory, which will be discussed thereafter. Divisibility rules, greatest common divisors, least common multiples, and modular arithmetic techniques are some of the topics that will be discussed during this course. In addition to the fact that these concepts are fascinating, they are also of tremendous aid in the process of finding solutions to a wide variety of issues that are demonstrated in the Math Olympiad.

The fundamentals of probability and statistics are covered in the second part.

In the area of unpredictability and chance, the ideas of probability and statistics are vitally significant aspects to take into account. Within this section, we will go over a few of the fundamental principles that underpin probability. Calculating probabilities, the principles of probability, and the fundamentals of combinatorics are all examples of notions that fall under this category. In addition, we will provide an introduction to core statistical concepts, such as the mean, median, mode, and range, which are required for the interpretation of data and the formulation of judgments. These ideas are essential for the data interpretation and judgment formulation processes. In the context of mathematical competitions, the ability to solve problems that include probability and statistics is an essential necessary talent to possess.

It is absolutely necessary to have a thorough understanding of ratios, proportions, and percentages in order to be able to solve the many various sorts of problems that are encountered in math competitions. You are going to be guided through the concepts of ratios and proportions in the following section, and you are going to learn how to solve problems that involve proportional relationships. Additionally, we will dive into the domain of percentages, covering subjects such as percentage rises and decreases, as well as the numerous applications of percentages in a number of scenarios. Aside from the fact that these concepts are essential to the study of mathematics, they are also of great assistance in events that take place in the real world.

A Brief Introduction to Set Theory (part 2.7) set theory

The topic of mathematical logic known as set theory focuses on sets, which are collections of objects. Sets are the subject of the study. Additionally, groups of goods are included in sets. This section will focus on the principles of set theory and will proceed to discuss them. The notion of sets, the many types of sets, and the methodology for describing them are some of the topics that will be discussed in this course. During the course of this discussion, all of the operations that may be performed on sets, such as union, intersection, difference, and complement, will be discussed. In order to make use of advanced mathematical reasoning and to find solutions to issues, it is necessary to have a good understanding of set theory.

In Section 2.8, we will be investigating sequences and series.

When it comes to the study of mathematics, the concepts of sequences and series are extremely important, particularly in the areas of number theory and algebra. The purpose of this section is to provide an introduction to arithmetic and geometric sequences, as well as the series that correspond to each subject. You will learn the knowledge necessary to discover the nth term of a sequence, to sum up the terms in a series, and to appreciate the concepts of convergence and divergence in series through the course that you are now enrolled in. It is necessary to have an understanding of these concepts in order to be able to recognize patterns and come up with answers to challenging problems when competing in mathematical competitions.

In the section titled "Graphs and Functions,"

A wide variety of mathematical subjects make considerable use of graphs and functions, in addition to being key components of algebra. Graphs and functions are also utilized extensively in algebra. As well as providing an introduction to the concept of functions, this part also discusses the various types of functions, including linear, quadratic, and exponential functions, as well as the graphing of these functions. It is vital to have the capacity to interpret and draw graphs in order to be able to provide a visual representation of mathematical concepts and to find answers to problems that include functional relationships.

Section 2.10 of the Basic Trigonometry Course

Despite the fact that trigonometry is generally considered to be a more difficult subject, it is vital to have a rudimentary understanding of the subject in order to compete in many mathematical competitions. Within this section, the principles of trigonometry are dissected and analyzed. The trigonometric ratios (sine, cosine, and tangent) and the ways in which triangles with right angles can be utilized to solve issues are included in these principles. By making use of these ratios, you will obtain the information that is necessary to solve problems involving angles and lengths in triangles. This is a skill that is essential in the process of addressing problems that involve geometry.

2

Strategies for Math Problem Solving

The skill of problem-solving, which is an essential talent in math Olympiads, is the subject of the second chapter, which shifts the focus for the second chapter. The knowledge that was taught in Chapter 1 serves as the foundation for this shift in focus, which builds upon that understanding. With the help of this chapter, students will be able to acquire the strategies, methods, and perspectives that are essential for successfully addressing challenging mathematical problems.

Methods for Solving Problems are Discussed in Section 3.1

The tasks that are part of the Mathematical Olympiad demand more than just a fundamental understanding of mathematics; they also require the ability to think strategically and approach problem-solving in a methodical manner from a mathematical perspective. We are going to look into a number of different methods for solving problems in this section. Some of these methods include working backwards, spotting patterns, simplifying complex circumstances, and employing logical reasoning. Students will be able to approach a wide array of issues in a more effective manner once they have gained a grasp of these tactics when they have completed their studies.

Effective Time Management During Examinations (3.2)

Within the realm of mathematics competitions, one of the most challenging challenges that can be encountered is the capacity to effectively manage time. In the next section, we will talk about many strategies that can be applied to successfully manage time while conducting examinations. This course will cover a variety of subjects, including the prioritization of activities, the allocation of time depending on the level of complexity and familiarity of the activity, and the determination of whether or not to move on from a problematic situation. In situations where time is a constraint, these strategies are an absolute requirement in order to attain the best potential level of performance in a competition.

Reasoning and Thinking Like a Logical Being are Required for 3.3

During math competitions, students are frequently put to the test in terms of their ability to think clearly and reason in a methodical manner. This section will focus on the process of building skills in logical reasoning, which will be discussed in the following section. We will discuss the value of deductive and inductive reasoning, as well as how to approach problems in a logical manner and how proofs are applied in mathematics, throughout this session. In addition, we will discuss how to approach problems in a manner that is logical. It is vitally necessary to enhance these talents in order to attain success in mathematical competitions and challenges that are of a higher level.

The Skills Required for Interpretation and Problem-Solving, Section 3.4

In math competitions, the objective is frequently not simply to find a solution to a problem; rather, it is to acquire an adequate understanding of the subject matter concerned. The purpose of this part is to explore the methods that can be utilized to analyze and comprehend mathematical information. During the course of this meeting, we will talk about different approaches of breaking down challenging situations into more manageable parts, gaining an understanding of the questions that are being asked, and locating information that is relevant. These capabilities are critically necessary in order to solve challenges in a manner that is both precise and effective.

3.5 An Overcoming of the Most Overcome Obstacles in the World

In the course of their participation in math Olympiads, students usually encounter a wide range of threats and challenges that are typical. In this section, we will discuss these challenges, which include how to deal with challenging circumstances, how to avoid making careless mistakes, and how to cope with the stress that comes with taking tests. For the duration of the competition, we provide direction and recommendations on how to overcome these challenges and maintain a positive attitude throughout the entire process.

Take use of the previous Olympiads to get experience with the problems.

A sufficient amount of practice is the key to successfully completing the questions in the Math Olympiad. This section includes a selection of problems from prior Olympiads, with the intention of providing students with practice opportunities. Not only are these difficulties supported by in-depth explanations and solutions, but they also provide insights into the thought process and strategies that were applied in order to overcome them. It is possible for students to better prepare themselves for Math Olympiads by practicing with these problems, which will help them become familiar with the structure of the issues as well as the level of difficulty that they include.

This section makes use of approaches that involve estimation and approximation.

There are a lot of issues that are presented at the Math Olympiad, and in many of them, precise calculations are not always required. In certain circumstances, it is even impossible to perform an accurate calculation. The art of estimating and approximation, which are skills that are absolutely required for quickly solving or checking issues, will be taught to you in this section. During this lesson, you will learn the processes for rounding numbers, estimating sums, products, and other operations, and making use of these estimates to limit down your options or evaluate the reasonableness of potential solutions.

Utilizing Algebraic Manipulation and Techniques as Part of Instructional Strategies

When it comes to providing solutions to a wide variety of problems associated with the Olympiad level, algebraic manipulation is a crucial component. In this section, more complex algebraic procedures are explained in greater detail than in the previous sections. Methods such as factoring, polynomial division, and the manipulation of complicated equations are included in this category of techniques. Students who are able to learn these ways can substantially boost their capacity to solve problems, which in turn helps them to come up with solutions that are both more efficient and elegant.

3.9 Techniques or Approaches for Working Through Geometry Issues or Problems

The difficulties that are presented in the geometry section of the Mathematical Olympiad can range from the most basic to the most challenging. This section discusses a variety of strategies that can be utilized while dealing with geometry challenges. The creation of exact diagrams, the utilization of geometric properties, and the application of theorems are all examples of these tactical approaches. To solve problems in an inventive manner, a special emphasis is placed on the capacity to perceive problems and the application of geometric intuition. This is done in order to solve problems.

The following are ten alternative approaches that can be utilized in order to tackle problems involving combinatorics and probability.

When it comes to the process of making decisions in the domains of combinatorics and probability, intuition and strategy are particularly important factors. A number of strategies for tackling combinatorics and probability problems are presented in this section, which serves as an introduction to those strategies. The use of counting procedures, a comprehension of permutations and combinations, and the use of probability laws are all examples of these approaches. As a result of the fact that students typically struggle with these topics, the objective of this part is to demystify them by presenting straightforward solutions and examples.

3.11 How the Learning of Mathematics Can Lead to the Development of Intuition

In order to be successful in Math Olympiads, it is very necessary to develop a strong mathematical intuition throughout one's life. Through the encouragement of inquiry, curiosity, and the identification of patterns, the objective of this section is to place an emphasis on the improvement of this intuitive ability. Methods for developing intuition include working on a range of problems, reflecting about alternative solutions, and obtaining an understanding of the core principles on which they are founded. This is in contrast to the traditional method of just memorizing formulas.

3.12 Making Preparations to Deal with Unanticipated Occurrences in Competitions

It is common knowledge that Math Olympiads will occasionally provide players with obstacles that they were not expecting. The students will be prepared for obstacles that they may encounter in the future that are unexpected or uncommon, which is the goal of this section. It explores strategies for keeping one's cool in high-pressure situations, thinking creatively, and tackling difficulties that are unfamiliar to oneself with a perspective that is both clear and sensible. The preparation that is being done is absolutely necessary in order to be able to adjust to any challenge that may be provided during a tournament.

3

Preparing for Specific Competitions

After covering the fundamentals of mathematics and problem-solving strategies, Chapter 3 focuses on preparing for specific math competitions. Each major competition has its own format, style, and types of questions. This chapter provides insights and preparation strategies tailored to each competition, helping students maximize their performance.

4.1 MathCON: Format and Strategies

MathCON is a popular national mathematics competition that attracts thousands of students each year. This section provides an overview of the MathCON format, including the types of questions and the scoring system. We will offer strategies tailored to the unique aspects of MathCON, focusing on how to approach its diverse range of problems, from multiple-choice to open-ended questions. Tips for effective preparation, including time management and problem-solving techniques specific to MathCON, are also discussed.

4.2 AMC-8: Tips and Practice

The American Mathematics Competitions (AMC) are a series of examinations and curriculum materials that build problem-solving skills and mathematical knowledge in middle and high school students. AMC-8 is designed for students in grade 8 and below. This section delves into the structure of AMC-8, covering the types of problems typically encountered and strategies for tackling them. We will also provide practice problems and tips on how to prepare effectively for AMC-8, focusing on building the necessary skills and confidence.

4.3 MATHCOUNTS: Preparation Guide

MATHCOUNTS is a highly prestigious national middle school mathematics competition. This section offers a comprehensive guide to preparing for MATHCOUNTS, including an overview of its format, which includes Sprint, Target, Team, and Countdown rounds. Strategies for each round are discussed, along with tips on teamwork and speed. Special emphasis is placed on the unique aspects of MATHCOUNTS, such as the fast-paced nature of some rounds and the importance of accuracy and efficiency.

4.4 Math Leagues: Understanding the Format

Math Leagues offer students the opportunity to compete at local, state, and national levels. This section provides an overview of the format of various Math League competitions, highlighting the differences and similarities among them. Strategies for preparing for Math League contests are outlined, focusing on the specific skills and knowledge needed to excel in these competitions. This includes problem-solving techniques, time management, and effective practice methods.

4.5 Preparing for International Mathematics Olympiad (IMO)

Although the International Mathematics Olympiad (IMO) is typically for high school students, advanced middle school students often show interest in the types of problems presented at this level. This section introduces the format and style of IMO problems, which are known for their complexity and creativity. We will discuss how to build the advanced problem-solving skills and mathematical reasoning required at this level, and how participating in or studying IMO problems can enrich a student's mathematical understanding.

4.6 Regional and State Competitions: Adapting to Various Formats

Apart from national and international competitions, many regions and states host their own math contests. This section focuses on how to adapt to the varying formats of these local competitions. It includes insights into understanding different scoring systems, problem types, and preparation strategies tailored to these contests. Emphasis is placed on the importance of understanding local curriculum standards, as many regional contests are aligned with them.

4.7 Online Math Competitions and Challenges

In today's digital age, numerous math competitions are held online, offering unique challenges and opportunities. This section covers how to prepare for and excel in online math competitions. Topics include navigating digital platforms, managing time effectively when competing online, and strategies for tackling the types of problems often found in online contests.

4.8 Special Focus: Team Competitions and Collaborative Problem Solving

Team competitions, such as those in certain rounds of MATHCOUNTS or in team-based math leagues, require a different set of skills, including collaboration and effective communication. This section offers strategies for excelling in team competitions, including how to work effectively with teammates, divide problems, and combine individual strengths for optimal team performance.

4.9 Utilizing Mock Competitions for Preparation

Mock competitions are an excellent way to simulate the actual competition environment. This section discusses how to organize and use mock competitions for effective preparation. It includes tips on creating a realistic competition experience, analyzing performance post-mock competition, and using these simulations to fine-tune strategies and skills.

4

Practice Problems and Solutions

An essential component of the book is Chapter 4, which provides a comprehensive compilation of practice problems together with in-depth explanations of every solution. It is the purpose of this chapter to provide practical experience with the different kinds of problems that are faced in Math Olympiads, as well as to reinforce the concepts and tactics that were discussed in the chapters that came before it.

Arithmetic Problems, Section 5.1

In this part, you will find a variety of arithmetic tasks, ranging from simple computations to more difficult word problems that involve percentages, ratios, and proportional reasoning. The purpose of the challenges is to improve individuals' computational abilities as well as their capacity to apply mathematical principles in a variety of settings. Solutions are offered together with detailed descriptions of each step, which ensures that a comprehensive comprehension of the procedures is achieved.

The Challenges of Algebra 5.2

The majority of math competitions include algebra as an essential component. Several different types of algebraic problems, such as linear equations, quadratic equations, and algebraic expressions, are presented in this section. There is a wide range of skill sets that can be accommodated by the challenges, which vary from simple to high levels. Providing significant insights into efficient algebraic thinking, the answers place an emphasis on algebraic procedures and problem-solving methodologies from a mathematical perspective.

Geometry Puzzles, Number 5.3

A wide variety of topics, including angles, triangles, circles, and polygons, are covered by the geometry puzzles that are included in this section. For the purpose of evaluating spatial reasoning as well as the application of geometric principles and theorems, these questions have been purposefully developed. Every solution is accompanied by illustrations that illustrate the answer, as well as a comprehensive explanation of the geometric ideas that are implicated.

5.4 Questions Regarding Number Theory

This part of the article goes into the fascinating realm of number theory, providing problems that pertain to prime numbers, divisibility, modular arithmetic, and other fundamental ideas. These challenges contribute to the development of a more profound comprehension of number theory and the applications of this theory. Clear explanations are provided by the solutions, and the logical rationale that behind each response is brought to light.

5.5 Exercises that Focus on Probability and Statistics

This area contains problems that are centered on probability and fundamental statistics. These problems include concepts such as probability rules, combinatorics, mean, median, mode, and range along with other related ideas. The purpose of these activities is to promote a deeper comprehension of statistical and probability concepts as well as their practical application. Students are better able to understand the subtleties of these issues when they are provided with solutions that explain the logic and calculations involved.

Problem Sets for Advanced Students

Students who are looking for a more difficult challenge will find that this part provides advanced problem sets that incorporate a variety of mathematical disciplines. These questions are comparable to those that are encountered in national and international math Olympiads, both in terms of their style and their level of difficulty. In addition to being comprehensive, the solutions offer insights into sophisticated problem-solving approaches and mathematical thinking.

5.7 Investigating Combinatorics and Mutations in Combinatorics

Students are given an introduction to the intriguing area of combinatorics and permutations in this part. Students are required to count, arrange, and analyze a variety of different combinations and sequences in order to complete the various challenges that are included in this package. These exercises are designed to help participants hone their skills in preparation for competitions, and they range from simple to more difficult settings. The solutions, which include the utilization of Pascal's Triangle, the counting principle, and factorial computations, place an emphasis on the fundamental ideas and tactics that underlie the problem.

5.8 Problems in Set Theory and Logic Analysis

Set theory and logic are fundamental aspects of higher-level mathematical thought that are necessary to understanding. An assortment of questions pertaining to sets, Venn diagrams, logical statements, and the applications of these concepts are presented in this area. The pupils are encouraged to think in an abstract manner and to reason rationally through these challenges. The solutions provide in-depth explanations and emphasize the significance of accurate thinking in the field of mathematics.

5.9 The Applications of Trigonometry and It Structure

Trigonometry problems are included in this part, which is intended for students who are prepared to move on to more challenging topics. The topics covered in these questions include trigonometric ratios and theorems, as well as their applications in a variety of geometric contexts. There are detailed explanations provided in the solutions of how to utilize trigonometric concepts in order to solve problems that are both theoretical and real-world in nature.

The 5.10 Word Problems and Their Applications in the Real World

The ability of pupils to apply mathematical concepts to real-world circumstances is evaluated through the use of word problems, which are a common component of math contests. Word problems involving a variety of mathematical topics are included in this part, which offers a wide collection of word problems. Students are led through the process of turning language into mathematical equations and effectively solving them through the use of solutions that are structured to guide them through the process.

5.11 Graph Theory and Problems Associated with Networks

In the field of mathematics, graph theory is an intriguing subfield that has practical applications in the fields of computer science, engineering, and the social sciences. The concepts of graphs, nodes, edges, and routes are presented in this part as an introduction to the fundamentals of graph theory. Students will be challenged to think about relationships and connections through the problems that are presented in this part. The solutions offer new perspectives on this segment of mathematics that has received relatively less attention but is becoming increasingly significant.

5.12 Riddles and Brainteasers in the Field of Mathematics

The inclusion of mathematical puzzles and brainteasers in this section is intended to add a sense of fun and creativity to the overall experience. Students are encouraged to approach mathematical problems from a variety of perspectives by being presented with these problems, which are designed to challenge their thinking. It is not enough to simply have the correct answer in order to solve these puzzles; one must also consider the creative process and the logical reasoning that was utilized in order to arrive at the solution.

Advanced Topics

Chapter 5 will expose students to more difficult subjects in mathematics, moving beyond the principles and problem-solving procedures that have been covered up until this point. The purpose of these themes is to present students with a mathematical challenge and to extend their grasp of mathematics. This is especially true for students who are aiming to achieve in the most difficult levels of mathematical contests.

The Investigation of Complicated Numbers

One of the most important aspects of advanced mathematics and engineering is the utilization of complex numbers, which represent a considerable expansion of the real numbers. In this part, the idea of complex numbers is presented, along with its algebraic form and the geometric meaning of their numerical values. The students will acquire the knowledge necessary to solve polynomial equations by performing operations with complex numbers and using those operations. Within the scope of this part, the applications of complex numbers to a variety of mathematical issues are also investigated.

6.2 Fundamentals of Trigonometry Introduction

As a continuation of the fundamental trigonometry that was presented in Chapter 4, this part goes even farther into the information. This course covers a variety of topics, including as advanced trigonometric identities, the unit circle, and the application of trigonometry to difficult difficulties. In addition, students will investigate finding solutions to trigonometric equations and applying trigonometry to the study of calculus.

6.3 Techniques for Advanced Algebra Studies

Systems of equations, inequalities, and functions are some of the more sophisticated subjects that are covered in this area of the algebra course. During this lesson, students will gain an understanding of radical equations, rational expressions, and polynomial functions. There is also a discussion of more advanced methods, such as the utilization of the discriminant, synthetic division, and the Remainder and Factor Theorems.

6.4 Constructions and Proofs of Geometrical Properties

The study of geometry is not limited to the resolution of problems; rather, it also involves the comprehension and demonstration of fundamental characteristics. Students are introduced to rigorous geometric proofs in this part, which focuses on geometric constructs that are made with a compass and a straightedge. Proving theorems regarding triangles, circles, and polygons are some of the topics that will be covered, along with investigating the notion of mathematical rigor in architecture and geometry.

Calculus, Section 6.5: An Introduction

This part offers an introduction to the fundamental ideas of differential and integral calculus from the perspective of students who are prepared to begin their journey into the world of calculus. Limits, derivatives, integrals, and the applications of these concepts are among the topics covered. Calculus is an important part of mathematics that is frequently used in higher education as well as in a variety of scientific domains. The purpose of this section is to provide students with a head start in grasping the fundamental ideas of calculus.

Mathematics of Discrete Forms 6.6

Discrete mathematics encompasses a broad variety of subjects, including logic, set theory, combinatorics, graph theory, and algorithms, amongst others. These fundamental ideas, which are essential in the field of computer science as well as other technical domains, are presented in this part. Students will be able to build a knowledge of mathematical reasoning and problem-solving in discrete mathematics more effectively with the aid of the problems and exercises contained in this part.

6

Tips From Former Winners

The purpose of this chapter is to assemble a collection of interviews, tips, and methods from Olympiad champions and high performers from the past. Not only does it offer students a source of inspiration and motivation, but it also offers a novel perspective on how to approach mathematical competitions.

7.1 Interviews and Additional Recommendations

The following section includes interviews with former winners of the Math Olympiad as well as participants who are recognized for their exceptional level of competition. In their conversation, they talk about the challenges they faced, the challenges they overcame, and the experiences they have had. In these interviews, which provide valuable insights into such elements, the mindset and strategies that lead to competitive achievement in mathematics are covered. These interviews include a wealth of information. The themes that are discussed include, but are not limited to, study habits, approaches to problem-solving, and ways for dealing with the pressure of competition.

7.2 Strategies and mentalities that can assist you in achieving victory

Math Olympiad winners who have previously won the tournament provide their perspectives on the strategies that contributed to their success in this particular section of the competition. They offer advice on how to deal with a variety of challenges, how to properly prepare for each one, and how to have a positive and productive attitude during the entire process. It is possible that students who are looking for strategies that have been tested and discovered to be effective and that can be applied in competitive contexts would find this area to be of great assistance.

To Get Ready for the Competition While Maintaining a Balance Between Schoolwork and Competition

Students who participate in Math Olympiads have a variety of challenges, one of which is the difficulty of balancing their preparation for the competition with their regular college education. This is one of the challenges that they face. When it comes to achieving a healthy balance between academic responsibilities and preparation for competitions, this section provides guidance on how to effectively manage time, how to develop priorities, and how to manage your time in general. In addition, there is instruction on how to deal with stress and maintain a healthy lifestyle while participating in the program.

7.4 The Long-Term Development of The Capabilities to Perform Mathematical Operations

The development of mathematical skills that can be utilized for an extended period of time is very important, and this is in addition to the preparation for specific events. This section will explore the ways in which participating in Math Olympiads may help one build a more thorough understanding of mathematics and increase problem-solving abilities, both of which are incredibly valuable in both academic and professional contexts. The objective of this section is to discuss the ways in which Math Olympiads can help one develop these talents. Former winners of math competitions share the manner in which the competitions have impacted their work careers and the intellectual pursuits they have followed in the past.

7.5 Inspiration & Motivational Inspiration: Inspirational

The students are the primary emphasis of this section, which is geared to providing them with an atmosphere of encouragement and motivation. In this discussion, former competitors of the Math Olympiad share their experiences of perseverance, the lessons they gained from both their successes and their failures, and the ways in which their involvement in math competitions has improved the quality of their lives. The objective of this section is to inspire students to follow their inclination for mathematics, irrespective of the outcomes of any competitions in which they might take part.

7.6 Confronting Challenging Circumstances and Learning from Your Errors: Taking on Complex Situations

In this section, we will address the relevance of the acceptance of challenges and the perception of setbacks as opportunities for personal development. Participants who have previously competed in the Olympics share the difficulties they have encountered and the ways in which those experiences have contributed to their growth and improvement. The importance of being resilient, the value of learning from one's mistakes, and the ways in which one may use failure as a stepping stone to reach success are all topics that are discussed in these articles.

Examining the Roles That Mentors and Coaches Play in the Process

There are a number of components that are incredibly crucial when it comes to preparing ready for Math Olympiads, including mentorship and coaching. Interviews with previous winners are included in the next section. These individuals share the impact that their mentors and coaches had on their personal development. They provide direction on how to select an appropriate mentor, the benefits of working together with a coach, and the methods in which to make the most of these relationships in order to enhance preparation for competitions for the purpose of improving performance.

7.8 The Constructing of a Community That Is a Comfortable Environment

When one goes through the process of competing in Math Olympiads, it is not just a journey that one takes on their own; it usually involves being a member of a community. The importance of forming a community of classmates, teachers, and family members who are supportive of one another is investigated in this portion of the article. In this section, prior winners explain the ways in which their community played a part in their accomplishment and provide suggestions on how to develop and sustain a collaborative network.

7.9 Methods for Making the Transition to Math Competitions Offered at Higher Levels of Competition

In this part, students who desire to continue competing in higher-level math competitions are provided with guidance on how to transition from more straightforward to more challenging arithmetic tournaments. Additionally, it includes advise on how to build upon the talents and experiences earned from middle school activities, as well as how to prepare for more difficult challenges, such as the International Mathematical Olympiad.

7.10 Mathematical Competitions and the Real World Outside of it

Taking into consideration the fact that the journey does not conclude with mathematical competitions, this section investigates life after these events have taken place once they have been completed. In this section, former winners share the ways in which their involvement in Math Olympiads has influenced their academic choices, professional paths, and personal growth based on the experiences they have gained. The participants investigate the transferable skills that may be acquired by participation in math competitions, as well as the manner in which these abilities can be utilized in a range of other domains.

7.11 Methods for Managing Time and Techniques for Studying

When it comes to achieving success in Math Olympiads, effective time management and study approaches are absolutely necessary. In this area, previous Olympians share their insights on how they successfully managed their study schedules, how they successfully combined their academics with their preparation for the Olympiad, and the strategies they utilized to make the most of their learning. In this section, you will find advice on how to create study programs, how to concentrate on challenging topics, and effective review procedures.

Understanding How to Handle Anxiety and Pressure 7.12

Taking part in Math Olympiads may be an experience that is fraught with intense pressure. This section discusses ways to cope with the stress and worry that are an inevitable part of being in a competitive environment. The previous champions discuss their individual methods for maintaining composure and concentration, which may include skills for mental preparation, mindfulness practices, and methods for maintaining motivation in the face of stress.

The Importance of Maintaining Both Physical and Mental Health 7.13

In the quest of academic brilliance, it is common to ignore the need of maintaining both one's physical and mental health. Olympiad veterans highlight the significance of maintaining a healthy lifestyle, which includes engaging in regular physical activity, consuming nutritious foods, and getting sufficient amounts of rest. They highlight the fact that it is possible for mental sharpness and general performance to be strongly impacted by one's physical well-being.

7.14 Taking Advantage of Individual Curiosity and Continual Education

Deep-seated curiosity and a passion for learning are two characteristics that are shared by successful Olympiad competitors across the world. In this part, we will discuss how to sustain and capitalize on this curiosity outside the realm of mathematical competitions. It provides guidance on how to pursue mathematical interests, investigate subjects that are connected to mathematics, and find methods to continue to be involved and challenged in mathematics.

7.15 The Importance of Critical Thinking and Personal Reflection

Seeking feedback and engaging in self-reflection are two common activities that are employed in the process of continuous growth in Math Olympiads. Within this part, we look into the ways in which prior champions improved their talents by utilizing feedback from coaches, peers, and their own personal thoughts. This article examines the significance of being receptive to constructive criticism as well as the function that self-reflection plays in the process of personal development and education.